T0130716

RAND

New-Concept Development

A Planning Approach for the 21st Century Air Force

Leslie Lewis, Zalmay M. Khalilzad,
C. Robert Roll

Prepared for the
United States Air Force

Project AIR FORCE

Preface

This research was undertaken at the request of the Air Force. The objective was to identify a process for developing new concepts, which is absent within the total Air Force. New ideas generated within the Major Commands (MAJCOMs) often have no place to be shared or adopted within the corporate Air Force.

The purpose of this work was to define what new-concept development is and to define a framework and process for new-concept development. The framework is designed to encompass all the steps in the new-concept development process. The proposed process links to the critical resource identification and allocation processes—requirements; the Planning, Programming, and Budgeting System (PPBS); and acquisition—and current Air Force planning initiatives.

This research was conducted as part of the Strategy and Doctrine Program.

Project AIR FORCE

Project AIR FORCE, a division of RAND, is the Air Force federally funded research and development center (FFRDC) for studies and analyses. It provides the Air Force with independent analyses of policy alternatives affecting the development, employment, combat readiness, and support of current and future aerospace forces. Research is being performed in three programs: Strategy and Doctrine, Force Modernization and Employment, and Resource Management and System Acquisition.

Contents

Figures

Tables

Summary

In early 1995, the Chief of Staff of the Air Force (CSAF) determined that the Air Force needed to strengthen its corporate planning capabilities for the near, mid-, and long terms. The planning function had to link strongly to the critical Department of Defense resource allocation and management processes.

Project AIR FORCE was asked to assist in defining a new-concept development framework and process that could support Air Force long-range planning. Concept development was to focus on the generation of new ideas and their incorporation into Air Force planning and programming activities.

Six issues were identified as critical to this analysis:

1. What is new-concept development?
2. Why is it important to the Air Force?
3. What are the elements of the process, and how do they interact?
4. How does new-concept development link to the Air Force's long-range planning activities and resource identification and allocation processes?
5. How should it be organizationally supported and nurtured?
6. How might the Air Force begin to institutionalize the process?

Concept development in the Air Force is the systematic, cooperative study and application of innovation. Innovation consists of new ideas, concepts, doctrine, devices (hardware), etc. Innovation can occur within any of the Title X functions or across the functions to improve the Air Force's overall ability to provide capabilities to the commanders in chief (CINCs).

New-concept development is important to the Air Force (or any of the services) because it has the potential to provide alternative concepts. Furthermore, new-concept development is essential to the Air Force's ability to accomplish military missions in a competitive environment.

The analytic framework used for this analysis was **demand**, **supply**, and **integration**. This framework provided a systematic way to identify and evaluate demanders of resources, their suppliers, and means of integrating the two elements. In planning for the far term, beyond the standard six to 15 years defense planners usually focus on, different elements shape demand, supply, and integration. For the outyears, the CINC requirements have not yet been defined, and the available resources are very uncertain. Thus, demand, supply, and integration are different in that they are not tied directly to national security objectives and fiscally constrained plans.

Defining Future Demand

Probable future demand is defined by three interdependent elements: future U.S. national security objectives and the challenges to them, the Air Force's vision, and its core competencies/capabilities. The current baseline provides a point of departure from which a range of future demands can be identified.

While the United States may not choose a single strategy, plausible alternatives must be assessed when determining what types of new concepts will be necessary in the future. The current and future strategic environments are marked by extreme uncertainty, which requires that flexibility be integral to the Air Force's planning and investment strategies.

The Air Force is currently involved in several internal assessments and redesigns of its vision: Global Reach, Global Power, Global Awareness. These activities are attempting to shape the future Air Force and its support of the joint environment. The goal of this work is to identify the essence of the Air Force and how it might change—to import ideas for providing the individual commander with a shared intent, framework for planning, guide for future acquisition, foundation for concept development, sense of organizational purpose, and sense of service corporateness. Although this report does not attempt to put forth a vision, recognizing the Air Force's current efforts in this area, that vision is an important element in shaping the demand for new concepts.

Core competencies/capabilities form the third element shaping the future demand.[1] This study suggests that, for the Air Force, a core competency is a robust capability. Robustness is what determines a service's dominance in a particular area. A service must have the skills and expertise that provide an important capability to claim a core competency. The capabilities that the services provide must be applicable across most mission scenarios, and the efficiency of the capability must be demonstrable. This research emphasizes that core competencies/capabilities must be defined within both a functional (as defined by Title X) and a joint context.

---

[1]In this report, "core competencies/capabilities" is understood to mean "core competencies and/or capabilities."

The core competencies/capabilities found in the Air Force Executive Guidance—air superiority and space superiority, global mobility, precision employment, and information dominance—were used for this discussion. The Air Force, like the other services, has been reluctant to reexamine and redefine its core capabilities to accommodate the increased focus on joint operations. This research, however, suggests that the joint environment creates a critical new marketplace in which the Air Force must compete.

To illustrate how core capabilities might be assessed within a joint context, the project team mapped the Air Force's core competencies/capabilities to an existing set of joint operational objectives and tasks. The Air Force was also evaluated against other services to show comparative advantages in certain joint operational objectives, and a set of sample screening criteria was used to illustrate how the Air Force might refine its core competencies/capabilities.

With these interrelated elements—the assessment of alternative futures, the use of Air Force vision, and the use of core competencies/capabilities—the Air Force can conduct a systematic analysis to delineate future areas in which it wants to hold dominance, while also looking at what courses of action are necessary to posture it for the future.

Defining the Supply

Historically, the Air Force has tended to focus on technological solutions. The current and projected fiscal defense environments suggest that the Air Force needs to integrate both materiel and nonmateriel solutions. These kinds of ideas make up the supply side of new-concept development.

Within the Air Force, several areas foster innovative ideas. These include the MAJCOMs and the Air University. The external community (including universities, governments, FFRDCs, and other military departments) is also a good source of new ideas and concepts. However, the external community has also been affected by changes in the defense budget, as well as by other outside pressures.

Since the early 1990s, in response to declines in the defense budget, the services have used research and development (R&D) dollars to sustain force structure and existing inventory. While the impact of this has not yet been fully examined, it is known that the services have had to seek nonmateriel solutions to meet their capability demands.

Ties between the federal government and universities in encouraging technological and scientific innovation have also fostered the education of scientists and technologists. But these organizations may be affected by the declines in defense expenditures, therefore making it difficult for them to sustain many of their scientific research programs. Additionally, in the last 20 years, U.S. predominance in technological innovation has been challenged. New global alliances have emerged, and private firms are exploiting R&D and innovation developed outside the firms. Research ties have also increased between universities and industry.

The Air Force needs to ensure that it can generate new ideas and concepts itself, without depending on other organizations. It also needs to continue to incorporate technological innovation, from both the inside and the outside, into planning and procurement. It must ensure that innovation includes nonmateriel solutions. Not only must the Air Force work on materiel and nonmateriel innovation, it must also reevaluate its organization. There is no central place where new ideas can be collected and assessed. Although the MAJCOMs identify operational needs, they are shaped and represented within the MAJCOM stovepipes, which hampers generation of larger concepts for the whole Air Force. The Air Staff and Secretariat staffs should be functionally and organizationally aligned to respond to the integration and application of ideas across the entire Air Force.

Integration and the Operational Thread

The integration process contributes to resource analysis and helps define the Air Force's strategic direction and investment strategies. New-concept development integration focuses on balancing an array of ideas (supply) against a number of demand elements (vision, alternative futures, and core competencies/capabilities). The result of this is refined core competencies/capabilities, alternative visions, feasible concepts, and selected strategies and investment decisions.

Some of the most critical elements for integrating new-concept development include articulating a vision and core competencies/capabilities in a joint environment; defining the current baseline; considering the impacts of the future U.S. strategy and security environment; determining potential new joint operational objectives and tasks; assessing new operational tasks against the current vision and investment strategies; identifying activities essential to the desired outcome; and developing proposed strategies.

New-concept development is a continuous, iterative process that links to all major Air Force activities. It also links to the PPBS, because new ideas and concepts must influence investment strategies. The strongest linkage between new-concept development and the PPBS process occurs during the planning phase. The process interacts with the Mission Area Planning and Functional Area Planning systems by formulating and defining new concepts to assist the MAJCOMs in their development of new operational concepts and the definition of future requirements. Furthermore, new-concept development can provide information to justify the Air Force's program to the Office of the Secretary of Defense, the Joint Staff, and Congress.

An illustrative operational thread shows the interaction among the different subelements, as well as interrelationships among demand, supply, and integration functions. The research team chose to expand on the space issue to provide an example of how new-concept development and its linkages to the Air Force's various resource decision processes enable an evaluation of applicability and potential utility to the Air Force.

Conclusions

Using the economic model of demand, supply, and integration, this report discusses the elements that shape the demand when attempting to define strategic direction and potential investment strategies in the 15- to 20-year time horizon. There is an emphasis on nonmateriel solutions in the supplying of new ideas, as well on allowing new concepts to be shared throughout the Air Force. The integration process filters new ideas against demand and enables the Air Force to link new concepts to resource investment processes, such as the PPBS.

Future work will address new-concept development's links to the planning and resourcing processes within the Air Force. It will examine

- How proposed new concepts might be identified as useful
- How new-concept development and long-range planning should be functionally and organizationally aligned
- How new-concept development and long-range planning can be implemented and sustained.

Acknowledgments

The project team would like to thank a number of people. We appreciate the support of Maj Gen John Gordon, Special Assistant to the Chief of Staff for Long-Range Planning, and Clark Murdock, Deputy Special Assistant to the Chief of Staff for Long-Range Planning. Maj Mace Carpenter (AF/XOXS) guided this work and provided invaluable comments. Our discussions on core competencies/capabilities versus core capabilities were of particular value. Col Tom Cedel, Air Force Revolutionary Planning Office, provided insight in all stages of the project. We would also like to thank Col Gene Collins (AF/XOXS) who supported the work and ensured that it was shared with the Air Force leadership. The late Maj Gen Robert Linhard (AF/LRP) agreed that a new-concept development capability should exist within the Air Force, iteratively challenged the findings of this study as they emerged, and always supported this work in its different phases. To him, we owe a special note of appreciation.

Glossary

AE	Acquisition executive
BMDO	Ballistic Missile Defense Office
BUR	Bottom-Up Review
C⁴I	Command, control, communications, computers and intelligence
CINC	Commander in chief
CJCS	Chairman of the Joint Chiefs of Staff
CPA	Chairman's Program Assessment
CPR	Chairman's Program Recommendations
CSAF	Chief of Staff, Air Force
DEPSECDEF	Deputy Secretary of Defense
DoD	Department of Defense
DPG	Defense Planning Guidance
FAPS	Functional Area Planning System
FFRDC	Federally funded research and development center
FOA	Field operating agency
FYDP	Future year defense plan
JOEPS	Joint Operations Evaluation Process
JROC	Joint Requirements Oversight Council
JS	Joint Staff
JSPS	Joint Strategic Planning System
JWCA	Joint Warfighting Capabilities Assessment
LRC	Lesser regional conflict
MAJCOM	Major Command
MAPS	Mission Area Planning System
MRC	Major regional conflict
NASA	National Aeronautics and Space Administration
NATO	North Atlantic Treaty Organization
NMD	National Missile Defense
OBPRM	Objectives-based planning resource methodology
OOTW	Operations other than war
OSD	Office of the Secretary of Defense
PAF	Project AIR FORCE
PEO	Program executive officer

PM	Program manager
POM	Program Objective Memorandum
PPBS	Planning, Programming, and Budgeting System
QDR	Quadrennial Defense Review
R&D	Research and development
SECDEF	Secretary of Defense
TENCAP	Tactical Exploitation of National Capabilities
TMD	Theater missile defense
UN	United Nations
USSOCOM	United States Special Operations Command
USSPACECOM	United States Space Command
VCJCS	Vice Chairman of the Joint Chiefs of Staff
WMD	Weapons of mass destruction
XOX	Air Force Plans and Operations, Strategic Planning Division

1. Introduction

In early 1995, the Chief of Staff of the Air Force (CSAF) determined that the Air Force needed to strengthen its corporate planning capabilities. The planning function had to link strongly to the critical Department of Defense (DoD) resource allocation and management processes, such as the Planning, Programming, and Budgeting System (PPBS),[1] the Joint Warfighting Capability Assessment and the Joint Requirements Oversight Council (JWCA/JROC),[2] the service requirements processes, and the acquisition processes.[3]

RAND was asked by the Air Force to assist in defining a new concept development framework and process that could support Air Force long-range planning. While long-range planning focused on defining a corporate vision and strategic planning (spanning 15 to 25 years), concept development was to focus on the generation of new ideas and their incorporation into Air Force planning and programming activities. The CSAF wanted to know how new ideas could be "grabbed" and incorporated into Air Force thinking. For instance, where might new ideas come from?

In response to these questions, RAND addressed how the new-concept development process supports Air Force planning. It also identified the various elements of new-concept development and proposed ideas for how the Air Force might proceed with institutionalizing the framework and process. Future work

---

[1]The PPBS is the process DoD uses to develop the program it presents to Congress. The process has three separate phases: Phase 1, planning, consists of the establishment of a fiscal top line and the general resource guidance to the individual services. The services, in turn, define their individual service planning guidance. Phase 2, programming, is the application of the fiscally constrained plans to the services' total requirements. The demands of individual services usually far outweigh the available resources; therefore, a service in this phase is forced to choose what it is going to fund. Phase 3 consists of budgeting the program defined in Phase 2. Phase 3 necessitates additional choices concerning when and how a program will be funded over the next two years.

[2]The JWCA is a relatively new process and is an outgrowth of the JROC's responsibilities to identify operational shortfalls to the commanders in chief (CINCs) and to determine which service proposals put forth could solve the shortfall. The JWCA was initiated approximately two years ago to assist the Vice Chairman of the Joint Chiefs of Staff (VCJCS) in the identification of the CINCs' priorities and their operational shortfalls for the near, mid-, and long terms. See Lewis et al. (1995).

[3]DoD uses the acquisition process to develop and field weapon systems. The process is multilayered, but basic oversight is provided by the Office of the Secretary of Defense (OSD) through the Deputy Secretary of Defense (DEPSECDEF). Each service is responsible for conducting its own acquisition program. The internal service management structures for major programs have three tiers: The civilian acquisition executive (AE) oversees all major programs and selected programs; portfolios of programs are usually managed by military program executive officers (PEOs), who oversee collections of individual programs, which are managed by program managers (PMs). The DoD acquisition process is defined and guided by several DoD instructions and directives.

will address in greater detail the institutionalization of the new-concept development process and its ties to the various planning and resource management processes within the Air Force.

This report discusses the elements of new-concept development, makes some suggestions for how the Air Force might organizationally and functionally support such an effort, and provides some top-level recommendations on how it might implement the process.

Six issues were identified as critical to this analysis:

1. What is new-concept development?
2. Why is it important to the Air Force?
3. What are the elements of the process, and how do they interact?
4. How does new-concept development link to the Air Force's long-range planning activities and resource identification and allocation processes?
5. How is it organizationally supported and nurtured?
6. How might the Air Force begin to institutionalize the process?

Approach

In defining new-concept development, the project team reviewed published materials on technological innovation, planning, core competencies/capabilities, and organizational and functional efficiency.[4] Interviews were conducted with individuals knowledgeable about innovation. Histories of prior attempts at innovation were consulted; for instance, the Army has an extensive history on the development and institutionalization of new-concept development. The DoD resource allocation and management processes were examined to determine how new-concept development might influence decisions and be accommodated by these processes.

This work also synthesizes and integrates some of the results from several parallel RAND research efforts: definitions of new-concept development, assessments of future security environments, and analysis of core competencies/capabilities.

Different pieces and elements of the framework and process were iteratively developed. They were fleshed out and debated with various RAND and Air

---

[4]In this report, "core competencies/capabilities" is understood to mean "core competencies and/or capabilities."

Force colleagues to define an end-to-end framework and process. The resulting framework and process are presented in this analysis.

Organization of This Report

This report contains six sections. Section 2 establishes the analytic framework. Section 3 assesses how the demands for new concepts are developed. Section 4 provides an assessment of where the supply of new ideas comes from and some of the potential difficulties the Air Force might have in developing nonmateriel solutions, which are an important element of new-concept development. Section 5 discusses the integration of the demand and supply elements of the process. The interactive nature of the process with other Air Force processes is demonstrated through the description of an operational thread. Section 6 summarizes our insights, makes some conclusions, and suggests additional areas for future work.

4

2. Analytic Framework

The work began with defining new-concept development and sharing those insights with the Air Force leadership.[1] There are many definitions of planning and organizational innovation, but none that specifically addresses new-concept development.[2] The reviewed definitions most often discuss planning and organizational change in service-based organizations; even discussions of technology-based organizations do not apply directly to the Air Force. They often focus on organizations with a single product, such as microchip development, rather than on large, complex technology-based organizations (such as the Air Force, which develops and deploys warfighting capabilities). For instance, the literature discusses core competencies as critical considerations in any long-range planning activity; they are generally defined as those activities that identify an organization's expertise and that provide it a comparative market edge over competing organizations (Smith, 1994).

Some published definitions of organizational innovation were adapted and applied to the Air Force.[3] The Air Force is a technology-driven organization; it provides airborne weapon systems and platform capabilities to warfighting CINCs. The services provide capabilities within the functions defined by congressional law. These are known as the Title X functions: (1) recruiting; (2) organizing; (3) supplying; (4) equipping (including research and development); (5) training; (6) servicing; (7) mobilizing; (8) demobilizing; (9) administering (including the morale and welfare of personnel); (10) maintaining; (11) the construction, outfitting, and repair of military equipment; and (12) the construction, maintenance, and repair of buildings, structures, and utilities.[4]

We concluded that **new-concept development is the systematic, comparative study and application of innovation.** Innovation can pose nonmateriel solutions

---

[1]The initial work by the project team is contained in a briefing given at the CSAF's Long-Range Planning Conference, 31 March 1995.

[2]More-recent studies that address this issue include Nelson (1993) and Quinn (1992).

[3]Innovation encompasses the methods and means by which companies master processes that are new to them. These processes are necessary if firms, or a service like the Air Force, are to remain competitive in industries where technological advance is imperative. In different contexts, staying competitive through innovation means different things. In one case, it may mean being at the forefront of technology. In another, it may be the adaptation of preexisting technology to local circumstances. Some common attributes of effective innovative performance that allow firms and organizations to master relevant technologies are competence in design and production, effective overall management, and the ability to assess consumer needs. (Nelson, 1993, pp. 4–5 and 508–509).

[4]Public Law 99-433, October 1, 1986.

to problems—new ideas, concepts, doctrine—and/or materiel solutions—devices (hardware) and systems. Innovation can occur within any of the Title X functions or across the functions to improve the Air Force's overall ability to provide capabilities to the CINCs. Innovation involves the work of many people related to the adoption of new inventions, ideas, concepts, etc. It involves the identification and iteration of ideas.

New-concept development, if institutionalized, can contribute to an organization's ability to compete in a number of areas. New-concept development can also assist in the nurturing and refinement of core competencies/capabilities. In the case of the Air Force, or any service, core competencies are core capabilities (sets of resources developed by the services) that are provided to CINCs in support of joint missions. Core capabilities need to provide the most efficient and cost-effective means of achieving operational objectives. Concept development, therefore, needs to be an iterative process that is permanent and enables an organization to incorporate new ideas into the development and refinement of its core capabilities. The process should enable innovative strategies and solutions to be introduced and to be linked to the Air Force's long-range planning activities and to influence its investment strategies.

New-concept development is important to the Air Force (or any of the services) because it has the potential to provide alternative concepts, which are essential for the identification of multiple planning and investment strategies, and, ultimately, for deciding on courses of action, all of which are critical to sustaining an organization.[5] For the Air Force, new-concept development is necessary because of the increasing uncertainties in the strategic, tactical, and fiscal environments. The Air Force, as all the military departments, is confronted with new and frequently ill-defined missions that involve nonlethal activities. The increased emphasis on joint operations also necessitates that the Air Force plan to support the joint commanders in a variety of joint operational tasks whose accomplishment is dependent on seamless interactions with the Army and the Navy. All these demands are occurring within a period of budget decline, which will not improve in the foreseeable future.[6] The synergism of these demands on the Air Force requires that it proactively plan and put forth new ideas—both materiel and nonmateriel.

---

[5]Planning development literature indicates that an organization's ability to generate and incorporate new ideas into its strategic planning is essential to its long-term survival. If an organization is not open to change, it runs the risk of losing its ability to compete. (See Sayles, 1993.)

[6]There has been a considerable amount of discussion within the DoD over the fiscal ceiling. The generally accepted view is that DoD budgets will either continue to decline or will remain stable without inflation added. Some analysts argue that defense expenditure will drop over the next several years in relation to the rest of the Gross National Product. (Williams, 1996.)

6

The project team utilized the RAND-developed analytic structure called **demand, supply, and integration.**[7] The methodology provided a systematic way by which demanders of resources, their suppliers, and the integration of the two elements could be identified and evaluated. Figure 2.1 shows the overall framework.

In the fiscally constrained planning environment (which covers approximately a ten-year period[8] in the DoD), demand, supply, and integration are relatively easy to define. As shown in Figure 2.2, the demand side is shaped by the national security objectives, national military objectives, and joint missions. The CINCs define the near-term requirements based on the resources needed to

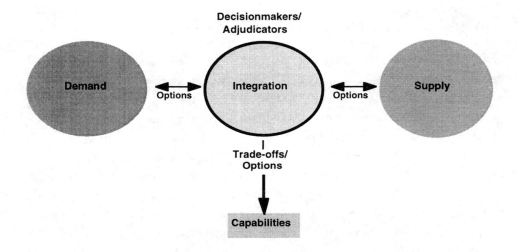

Figure 2.1—The Framework

[7]This analytic structure has been used in several RAND organizational and functional analyses. (See Lewis, Coggin, and Roll, 1994.)

[8]Some would argue that fiscally constrained planning covers only the six years defined in the Future Year Defense Plan (FYDP); however, most DoD programs, such as modernization, procurement, and even force structure allocations, span up to 15 years. Therefore, long-range planning must not only influence the six- to 15-year period, but also look beyond it.

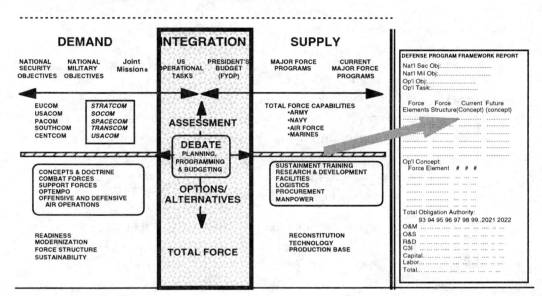

Figure 2.2—Long-Range Planning and the Operationally Based PPBS

perform their missions and the associated tasks. They must be operationally ready, be modernized, have sufficient force structure, and be able to sustain their operations.

The supply (or available resources) is provided by the total force capabilities put forth by each of the military departments in their provider roles. A critical part of the military departments' "provide" activities is to anticipate what future warfighting requirements might be. This activity is important, given that the services are the providers of capabilities and, therefore, must look beyond the current investment years and anticipate what the operational CINCs might need to perform a variety of missions, some of which have yet to be formally defined. Again, proposed solutions can be materiel and nonmateriel.

The integration of demand and supply is done by OSD and the Chairman, Joint Chief of Staff (CJCS). Congress directs the CJCS (and by default the Joint Staff [JS]) to integrate, determine the priorities, and to represent the CINCs' requirements in the DoD PPBS and in the requirements process. The CJCS and the VCJCS do this through the JWCA, which identifies CINC requirements and priorities. The JWCA's findings are iteratively presented to the JROC, whose purpose is to identify capability shortfalls.[9]

The military departments' resource priorities and decisions are contained in their Program Objective Memoranda (POMs), which are incorporated into the DoD's program, which is presented to Congress for funding.

When planning for the far term, beyond the standard six to 15 years defense planners usually focus on, different elements shape demand, supply, and integration. For the outyears, the CINC requirements are not defined, and the available resources are very uncertain. Thus, demand, supply, and integration are different in that they are not tied directly to national security objectives and fiscally constrained plans (i.e., they are demand driven). Therefore, different elements have to be applied in each of the areas.

The long-range planning and innovation literature suggests that several functional elements have to be present when trying to establish a long-range planning capability within a corporation. Two of the most critical elements are (1) a well-articulated corporate vision that looks out 10 to 15 years and (2) a set of core competencies/capabilities that are understood by the corporation's workforce. The two elements are interrelated. For instance, an altered corporate vision could lead to redefining or even discarding (over time) a core competency/capability that is judged by the corporate leadership no longer to provide a major competitive advantage. Conversely, the reinvigoration of a core competency/capability could contribute to changes in the corporate vision. Both of these elements, however, are shaped in the corporate world by an ongoing evaluation of the marketplace.[10] The marketplace shapes a corporation's vision and ultimately defines its core competencies/capabilities.

The military departments, however, are not corporations. They are directed as to what capabilities they are to provide to support the national security objectives. Furthermore, their core competencies/capabilities are often not unique to their organizations. For instance, the Air Force, Navy, and Army all have robust

---

[9]The JWCA is divided into ten categories: Strike; Land and Littoral Warfare; Strategic Mobility and Sustainability; Sea, Air, and Space Superiority; Deter/Counter Proliferation of Weapons of Mass Destruction (WMD); Command and Control; Information Warfare; Intelligence, Surveillance, and Reconnaissance; Regional Engagement/Presence; and Joint Readiness. Until recently, the JROC's purpose was to function as a review of service system proposals. Within approximately the last two years, this has changed to the identification of CINC operational shortfalls and resource alternatives. Part of the JWCA's purpose is not only to identify current capability shortfalls and posit investment alternatives but to seek out new solutions—materiel and non-materiel—by which to overcome the identified shortfalls.

[10]The Air Force is tested in a number of marketplaces, although these markets are very different from those of corporations. These include: (1) the budgeting market (PPBS, authorizations, and appropriations), (2) the market that assigns missions to services, and (3) the marketplace of war and operations other than war (OOTW), which determines ultimate success or failure.

capabilities in the area of information warfare, each pointing to these capabilities as critical to its operations. There is competition among the services to have a comparative advantage in a capability area; for instance, the Navy provides sea-based capabilities, the Air Force provides air and space capabilities, and the Army provides ground capabilities, but in an increasingly joint environment, the application of these capabilities is no longer clearly delineated.

The project team, therefore, concluded that the demand for new concepts emerged from three interrelated elements: (1) an assessment of the U.S. national security environment and the external security environment that looks out approximately 25 years; (2) a corporate vision of how the Air Force is going to support the U.S. national security objectives through its Global Reach, Global Power, Global Awareness; and (3) its core competencies/capabilities.

The assessments of the future demand provide the mechanism to identify those areas for which new concepts and ideas are necessary. The new concepts and ideas form the supply; they are the selected areas for future development in response to future demands. The integration function provides the balancing of the demand and supply. It defines when—near, mid-, or long term—some or all of a concept might be incorporated into the Air Force. It leads to redefining a strategic direction and the supporting investment strategies. It, therefore, must link back to the organization's vision and core competencies/capabilities. For instance, a new concept that has been accepted by the Air Force leadership as a worthwhile area for investment might contribute to a refinement of both the Air Force vision and a particular core capability. Figure 2.3 shows the analytic framework for new-concept development; the following three sections define in greater detail demand, supply, and integration.

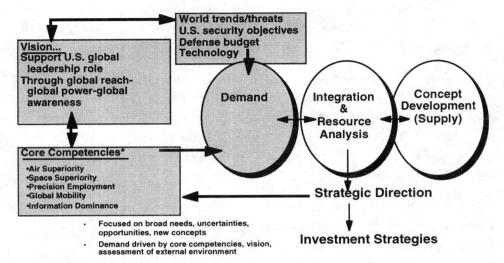

Figure 2.3—Planning for the Far Term

3. Defining the Future Demand

The future demand is defined by three interdependent elements: (1) U.S. security objectives and the future security environment, (2) the Air Force's vision, and (3) its core competencies/capabilities. The baseline for defining the spectrum of future demand begins by identifying the current environment in which the Air Force operates. The current baseline provides a point of departure from which future demands can be identified and assessed.

Future Security Environment

The current strategic environment is typified by the emergence of a wide variety of threats. In the last five years, the United States has been involved in a diverse set of missions—Operations Desert Shield and Desert Storm (Persian Gulf), Operation Restore Hope (Somalia), Operation Restore Democracy (Haiti), and Operation Joint Endeavor (peace enforcement in Bosnia). DoD planning, however, is shaped by several activities and the fiscal top line.[1] Although planning documents give a nod to increased threats from the proliferation of WMD and lesser regional conflicts (LRCs), the DoD resource focus remains on providing capabilities for two nearly simultaneous major regional conflicts (MRCs). Indicators are that the new strategy and resource review, called the Quadrennial Defense Review (QDR), will address a broad range of missions and not just the resources needed to support MRCs.

The United States has responded to the increased demands to perform OOTW.[2] Although the President has declared that the United States must sustain its ability to act unilaterally to protect U.S. national interests, the country is increasingly finding itself involved in missions that necessitate cooperative efforts with other nations and international organizations, such as the United Nations (UN) and the North Atlantic Treaty Organization (NATO). Recently, the

---

[1]In 1992, there was the Bottom-Up Review (BUR). The BUR focused on what capabilities are needed in a post–Cold War environment. The document's purpose was to shape proactively the DoD environment beyond the POM years by providing a broad scope of needs and potential investment areas. The document identifies a number of new threat areas such as WMD, but its planning assumptions are that the U.S. military must be able to support two simultaneous MRCs. (Aspin, 1993.) In 1994 and 1995, there was the congressionally mandated Commission on the Roles and Missions of the Armed Forces (CORM). It did not move away from two near-simultaneous MRCs. In 1996 and 1997, the QDR will address capabilities and future missions.

[2]OOTW includes a wide span of mission areas, such as peacekeeping, humanitarian assistance, peace enforcement, and infrastructure development.

OSD indicated that, during the QDR, the current baseline will be reevaluated, and attempts will be made to shift the national military strategy's focus from a force structure–based resourcing strategy to one based on capabilities. This redirection is in response to increased demand from the international community for U.S. support of non-warfighting missions and to projected declines in defense expenditures. The decline in DoD resources probably will not abate in the near term. Defense expenditure is currently at 250 billion constant dollars for FY96, with all projections showing a downward spiral through 2003. The Air Force will be funded at $60.3 billion in FY98 and at $70.7 billion for FY03.

The institutionalization of jointly supported missions also offers significant challenges to the Air Force. The JWCA and JROC processes have broadened the CJCS's charter in that he is now involved not only in setting priorities on the CINCs' requirements but also in defining the operational shortfalls and the capabilities needed to overcome these shortfalls. The Chairman's Program Recommendations (CPR) document identifies operational shortfalls, new missions that are emerging for the CINCs, and the joint capabilities that are necessary to overcome the identified deficiencies. The Chairman's Program Assessment (CPA) assesses the services' response (contained in their POMs) to identified joint operational shortfalls and suggests capability and investment alternatives.

During the Cold War, U.S. defense planning revolved around a single strategy— containment. The collapse of the Soviet Union left the United States without a grand strategy for how to operate in the post–Cold War environment. Although lacking a grand strategy, the United States has a national military strategy that focuses on regional threats, particularly in Korea and the Persian Gulf, with the capability to fight two MRCs nearly simultaneously. Technological changes continue to pose new threats to U.S. security.[3]

The United States will likely seek to maintain a position of global leadership, thereby precluding both the rise of another global rival and multipolarity. Among the likely requirements for realizing this overall objective are to

- Maintain and selectively extend the network of alliances and cooperation among the economically most-capable democratic nations

- Preclude hostile hegemony over critical regions

- Hedge against Russian reimperialization and Chinese expansionism while promoting cooperation with both

---

[3]This alternative futures section is extrapolated from Khalilzad (1996).

- Preserve U.S. military preeminence by maintaining the right force size and mix

- Maintain U.S. economic strength and an open international economic system and reduce the social crisis in our country

- Be judicious in the use of force, avoid overextension, and achieve effective burden-sharing with allies

- Obtain and maintain domestic support for global leadership and a strategy able to support it.

The United States could choose any one of these strategies in response to various global trends. Or it might choose to adopt none of these strategies and operate on a case-by-case basis. The key point is that each plausible alternative generates needs for new concepts that enable the military to function effectively within that set of scenarios. Whatever strategy the United States chooses to pursue, several trends suggest that the strategic environment in the near and midterm will be uncertain. To summarize a few of these trends,

- In many regions, there is an identifiable trend toward democratic governments and free markets.

- The proliferation of WMD could have a profound effect on future U.S. missions and investment strategies in national missile defense (NMD) and theater missile defense (TMD) systems.

- China is undergoing extremely rapid and revolutionary change along every dimension of national power.

- Europe's future security will depend on what happens in Russia, East Asia, Europe, the Balkans, and the Middle East. The current NATO operations in Bosnia will have a lasting effect on U.S. relations with its NATO allies. If the Bosnian mission fails, it could threaten NATO's existence.

- The United States could be asked to play an increased role in peacekeeping and peace enforcement operations.

- There are new vulnerabilities because of changes in technology and concept of operations, such as the vulnerability of our military forces and society to information attacks.

Depending on what strategies the United States employs, it may respond differently to these challenges in the future than it would in 1996.

14

Air Force Vision

The Air Force is currently involved in several internal assessments and redesigns of its vision: Global Reach, Global Power, Global Awareness. These activities are attempting to shape the future Air Force and its support of the joint environment, including both MRCs and OOTW. The goal of the vision work is to identify what the essence of the Air Force is and how it might proceed. The work is also attempting to develop more coherence for the Air Force by clearly articulating its purpose and the theory, doctrine, and strategy for accomplishing the Air Force's missions.[4] A shared vision provides a commander's intent, a framework for planning, a guide for future acquisition, a foundation for a concept development, a sense of organizational purpose, and service corporateness.[5]

The CSAF argues that a vision is critical for the Air Force because it provides clear goals and objectives for a diverse organization, which increases the organization's coherency to deal with complex issues. It also provides a common thread for the critical resource identification and allocation processes—PPBS, policy and doctrine development, MAPS, JROC/JWCA, Joint Strategic Planning System (JSPS)/Joint Operations Evaluation Process (JOEPS), and long-range planning.[6]

This report does not attempt to put forth a vision. It recognizes the Air Force's current efforts in this area. However, the vision work, as CSAF has noted, is an important element in shaping the demand for new-concept development.

Core Competencies

Core competencies form the third element shaping the future demand. Current literature on corporate reengineering discusses core competencies and their importance in defining and shaping an organization's vision and strategic planning. The literature consistently argues that core competencies can increase the focus of an organization by enabling it to concentrate on activities such as investment, research, manufacturing capabilities, and technological innovation. **A core competency is defined as an activity or product that enables a corporation to stand out as a world-class competitor in a particular area** (Quinn, 1992). This study has concluded that, for the Air Force, a core competency is a robust capability. As noted earlier, the military departments might all share a similar capability, but a particular service's ability to provide a capability with an

---

[4]CSAF Long-Range Planning Conference, 31 March 1995.
[5]CSAF Long-Range Planning Conference, 31 March 1995.
[6]CSAF Long-Range Planning Conference, 31 March 1995.

operational edge is what determines a service's dominance in a particular area. Furthermore, a dominant capability is more than the mastery of a single technology; it can include operational concepts. The Joint Staff is increasingly focusing on nonmateriel solutions to augmenting U.S. military capabilities while reducing costs.

As in corporations, once a service core capability is developed, it may even assist in the redefinition of the organization's activities. For example, the Air Force's recognized technological and operational mastery of heavy launch capabilities for space has facilitated its inclusion of space over the last 20 years as one of its key core capabilities. Its desire to shape and dominate space activities has led to major shifts in its vision and investment strategies. Innovation supports and refreshes the iterative discussion of an organization's core capabilities, for it can facilitate the development of new ideas and strategies. Conversely, the incorporation of new technologies and operational concepts has been critical to the Air Force's development and sustainment of its predominant role in the space debate.

For the services, core capabilities must contain several attributes. To claim a core capability, a service must have the skills and expertise that provide an important national security capability. The skills and capabilities must create and maintain real distinctions among the services; they must also be critical to the achievement of a strategic concept. The capabilities that a service provides must also be important in the future. The capabilities must be applicable across most mission scenarios, and their utility and efficiency must be demonstrable. Some examples of demonstrable effectiveness measures might be faster, more decisively, less risk, less collateral effects, and fewer forces needed. Core capabilities must also enhance an organization's competitiveness in the future. Therefore, to own a core capability, a service must be a key player in the critical strategic decisions that affect that capability, even though other services might also have interest and investment in that capability area. For instance, although the Navy has a substantial aviation capability, the key strategic decisions regarding air power reside with the Air Force.

Within the last five years, the services have increasingly attempted to justify their programs and long-term investment strategies within a joint context, in response to OSD guidance and Joint Staff activities, such as the BUR and JWCA/JROC processes.[7] The services, however, have been reluctant to reexamine and

---

[7]In 1986, the Congress passed the historic Goldwater-Nichols legislation that increased the power of the CJCS. The legislation, as noted earlier, has empowered the CJCS to define, evaluate, and contribute to the DoD's planning, programming, and budgeting activities to ensure that the CINCs' requirements are being sufficiently addressed within a joint context.

redefine, if necessary, their core capabilities to accommodate the increased focus on joint operations. joint issues are usually considered after the service POMs are completed; they are often included as addendums to the POMs rather than as integral to POM considerations. This research, however, suggests that the joint environment creates a critical new marketplace in which the Air Force must compete and, therefore, necessitates that any consideration of core capabilities be done, in part, within the context of joint operational objectives and tasks.

The debate within the Air Force over what constitutes a core competency/capability has yet to be settled. Over the last two years, the Air Force leadership has rigorously debated its core competencies/capabilities and their linkage to the Air Force's vision. All the participants agree in principle as to the importance of core competencies/capabilities and their ties to the Air Force's vision; the dilemma is what should be included in the list of core capabilities. This research does not address and does not attempt to resolve this debate but rather emphasizes that core competencies/capabilities are critical to shaping the demand side of new-concept development. They, therefore, must be defined within both a functional (as defined by Title X legislation) and a joint context. They must also enable the service to link them to joint tasks and facilitate the development of different concepts of operation. For this discussion, we used the list and definitions of core competencies/capabilities found in the Air Force Executive Guidance[8]:

- **Air Superiority and Space Superiority** are the degree of control necessary in air and space to position, maneuver, employ, and engage with forces of all media, while denying the same ability to adversary forces.

- **Global Mobility** is the timely positioning of forces through air and space, across the range of military operations.

- **Precision Employment** is the Air Force's ability to employ forces precisely against an adversary to degrade his capability and will, or the employment of forces to effect an event across the spectrum of conflict.

- **Information Dominance** is the ability to collect, control, exploit, and defend information while denying an adversary the ability to do the same.

The potential difficulty with this list of core competencies/capabilities and their definition is that they do not translate into specific joint operational objectives or

---

[8]The Air Force Executive Guidance is an internal document published by Air Force Plans and Operations, Strategic Planning Division. The document contains a wide variety of ideas and attempts to define the Air Force leadership's priorities and assumptions that underpin the Air Force of the future. Air Force Guidance, 13 September 1995.

tasks; they also do not provide sufficient guidance for future concept development or investment. They provide no insights as to what is important to the future Air Force. And they have not been translated into specific capabilities. For instance, which of these competency/capability areas is most important to the Air Force? In each of these areas, which joint activities/tasks does the Air Force want to dominate? What is its comparative advantage now in these activity/task areas, and what new concepts might be necessary to ensure continued or future dominance?

Critical to the identification and justification of core capabilities is the ability of the Air Force to discuss and justify them within a joint context. The Air Force's core competencies/capabilities were mapped to an existing set of joint operational objectives and example tasks. The mapping was not meant to be definitive; rather, it was done to illustrate how the Air Force might assess its core competencies/capabilities within a joint context. The illustrative joint operational objectives were extracted from the RAND-developed Objectives-Based Planning Resource Methodology (OBPRM).[9] Operational objectives define the goals of a particular operational activity. The example tasks are the activities that must be performed to accomplish a particular objective. The Air Force's core capabilities are linked to the objectives and their associated tasks. Figure 3.1 shows an illustrative mapping. For instance, our preliminary assessment indicates that the core competency/capability, air superiority, links to the critical joint operational objectives Shaping the Environment, Deter Aggression and Prevent Conflict, and Deploy Combat-Ready Forces. Each of the tasks associated with the operational objectives must be assessed according to the core capability and its ability to provide resources and operational concepts to efficiently accomplish a joint task.

Again, as a way to illustrate how the Air Force might link operational objectives to its core competencies/capabilities and to resources, a sample assessment was done. Table 3.1 shows a preliminary and qualitative assessment of the Air Force's comparative advantage against the other services to perform certain joint operational objectives and their associated tasks. If the Air Force was serious

---

[9]Since Objectives-Based Planning (OBP) is used so widely in the Pentagon, RAND, in conjunction with its sponsors, has pursued a common taxonomy for describing missions, objectives, and tasks. The result of the most recent review is described in Pirnie (1996). Project personnel from this study and related PAF activities participated in the review. Joint operational objectives, as used in this analysis, are now incorporated in an updated set of missions of the combatant commanders and supporting operational objectives. The Air Force uses OBP in its MAPS process; there it is called Strategy-to-Tasks.

18

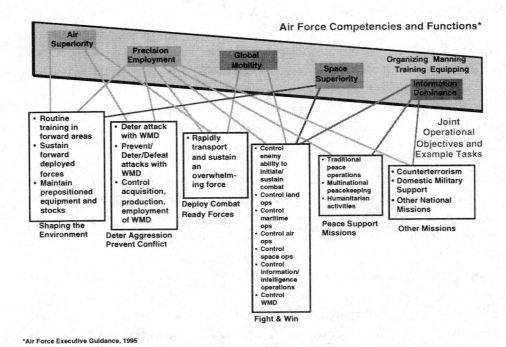

Air Force Competencies and Functions*

Air Superiority

Precision Employment

Global Mobility

Space Superiority

Organizing Manning Training Equipping

Information Dominance

Joint Operational Objectives and Example Tasks

- Routine training in forward areas
- Sustain forward deployed forces
- Maintain prepositioned equipment and stocks

Shaping the Environment

- Deter attack with WMD
- Prevent/ Deter/Defeat attacks with WMD
- Control acquisition, production, employment of WMD

Deter Aggression Prevent Conflict

- Rapidly transport and sustain an overwhelming force

Deploy Combat Ready Forces

- Control enemy ability to initiate/ sustain combat
- Control land ops
- Control maritime ops
- Control air ops
- Control space ops
- Control information/ intelligence operations
- Control WMD

Fight & Win

- Traditional peace operations
- Multinational peacekeeping
- Humanitarian activities

Peace Support Missions

- Counterterrorism
- Domestic Military Support
- Other National Missions

Other Missions

*Air Force Executive Guidance, 1995

Figure 3.1—Example of Matching Core Competencies with Joint Operational Objectives

about doing such an analysis as a way to refine its core competencies/ capabilities, the process would need to be objective and more quantitative. The × symbol (in Table 3.1) shows that a service is perceived to be the leader in a particular area. Listed at the top of the figure (moving left to right) are the Air Force, Army, Navy, Marines, and U.S. Special Operations Command (USSOCOM). USSOCOM is listed because it competes among the services as a provider of capabilities to perform certain joint tasks. Significantly, this figure underlines several points made earlier in this discussion. The Air Force's core competencies/capabilities must compete among those of the other services. In many areas, the services share a particular dominance of a task because they each contribute something unique to the performance of operational concepts to accomplish total joint tasks; on occasion, however, the capabilities are unnecessarily duplicative, leaving advocacy open to OSD and CJCS scrutiny. It is in these instances that an examination of the capability and its justification for supporting a core competency/capability might be required to determine whether this is a key area in which a service should be investing to protect and promote what it views as its vital institutional interests.

Table 3.1

Relative Contribution of the Services to Joint Operational Objectives

X = Perceived Leader
X = Contributor

Joint Operational Objectives	AF	Army	Navy	Marines	SOCOM
Shaping the Environment					
Routine training in forward areas	x	x	x	x	x
Sustain forward deployed forces	**X**	**X**	x	x	x
Maintain prepositioned equipment & stocks		**X**		**X**	
Deter Aggression & Prevent Conflict					
Prevent/Deter/Defeat Attacks with WMD	**X**	x	**X**		x
Control acquisition and production of WMD	**X**		x		x
Deploy Combat Ready Forces					
Rapidly transport & sustain an overwhelming force	**X**	**X**	x	x	x
Fight and Win					
Control enemy ability to initiate/sustain combat	**X**	x	x	x	
Control land operations	x	**X**		x	
Control maritime operations	x		**X**	x	
Control air operations	**X**	x	**X**	x	x
Control space operations	**X**	x	x		
Control information/intelligence operations	x	x	x	x	x
Control WMD	x		x		x
Peace Support Missions					
Traditional peace operations	x	**X**	x	x	x
Multinational peacekeeping	**X**	**X**	x	x	x
Humanitarian activities	x		x	x	**X**
Other	**X**	**X**			
Other Missions					
Counterterrorism					
Domestic Military Support	x	x	x		**X**
Other National Missions	x	x			

Conversely, a service's assessment of its ability to perform joint operational tasks can also contribute to the refinement of particular core competencies. Table 3.2 illustrates how this might be done. The table shows a set of sample screening criteria. These are drawn from the published literature on how service and technology corporations assess what core capabilities they should be maintaining, refining, or eliminating (see Smith, 1994). Some explanation of the criteria is useful. The first area identifies **skills that provide an important national security capability.** This particular criterion has been derived from the business literature (see Quinn, 1992), which notes that a corporation needs to identify core competencies/capabilities that fill a particular niche in the external marketplace. We have further refined this category by also noting that the core competency/capability must create and maintain a real (not an internally perceived) distinction among the competitors. And in the case of the services, it must also be critical to the development of a strategic concept and ultimately its

Table 3.2

Identifying Core Capabilities Utilizing Joint Operational Tasks

Screening Criteria Sample Joint Operational Tasks	Skills that provide an important national security capability • Essential to National Military Strategy • Critical to a strategic concept	Expertise that is recognized externally to provide a critical capability •Create and maintain real and meaningful distinctiveness •Critical to a strategic concept	Sets of capabilities that will be important in the future • Across all scenarios • Faster, more decisively, less risk, less collateral effects, fewer forces needed
Identify and monitor WMD facilities of potential adversaries			
Monitor WMD proliferation	Provide reconnaissance Ability to detect HEU		
Establish and defend safe areas	Provide reconnaissance Position satellites		
Gain air supremacy	Air Force - space element		
Counter enemy theater ballistic missiles	Army, Air Force, Navy - TMD		
Sustain U.S. space operations	Provide launch and positioning Space architecture	Air Force - lift Army - TENCAP	. . .
Render humanitarian assistance	Air Force - lift Army - ground support	Medical, water, etc.	Lift to get equipment to site
etc.			

operational execution. The first criterion is further refined in the second column by indicating that a service must provide the **expertise that is recognized externally to provide a critical capability.** This guideline underlines a point made earlier that, although all the services can possess the skills to provide a capability, a service must also be externally recognized as owning the expertise to be involved in the strategic decisionmaking in a particular capability area. And, finally, the service must possess the sets of interrelated capabilities that will be important in the future. The capabilities must either be unique to the performance of a task or be applicable across many scenarios.

The left-hand side of the figure lists a series of sample joint operational tasks drawn from the RAND Objectives-Based Planning methodology. The tasks shown are associated with the operational objectives of Shaping the Environment, Fight and Win, Peace Support Mission, and Other Missions. The illustrative assessment focuses on the Air Force core capability of **space superiority.** The partially filled-in evaluation shows that, in several task areas— Monitor WMD and Establish and Defend Safe Areas—the Air Force possesses critical skills and some expertise. For instance, in the latter task, the Air Force provides reconnaissance and positions satellites, which are essential task elements and support the core competency/capability of space superiority. In other task areas, however, the other services are also critical capability providers. For example, in the Counter Enemy Theater Ballistic Missiles task, the Army, Navy, and Air Force all participate in providing TMD capabilities. No single service, however, is viewed as possessing absolute expertise in this area; therefore, strategic decisionmaking is shared by all three. Again, these are illustrative of the types of analysis that must occur in the Air Force to ensure that its core capabilities are sufficiently responsive to the current and future joint environment.

A systematic analysis of core competencies/capabilities can contribute to the Air Force beginning to delineate future areas in which it might want to hold dominance and what courses of action—vision, investment strategies, new concepts—might be necessary to posture it for the future. Linking of tasks to current and future core capabilities is the first step in defining that process. Two illustrative examples of how the linkages and, ultimately, the courses of action defined are drawn from the application of this framework follow:

1. The current U.S. strategy requires **global mobility**. Airlift is a critical element of global mobility, and there is a need to provide both inter- and intratheater capabilities to support this demand. Airlift could be outsourced to commercial airlines, and the Air Force could still retain its strategic decisionmaking role in defining airlift, because the Joint Staff and OSD both concur that the Air Force provides this capability. However, in hostile environments, the safety of commercial airlift could be questionable; thus, military airlift must also be provided. Global mobility and airlift is an Air Force core capability. Substantial airlift could be outsourced in nonhostile environments. Cost-effectiveness issues surrounding military-provided airlift—training, procurement, and maintenance—need to be addressed. The Air Force will address these issues and will continue to shape and define the future airlift capabilities in response to CINC demands. These activities are essential for the Air Force to retain its competitive edge in providing global mobility.

2. The current U.S. strategy requires space-based capabilities to support an array of intelligence and information demands. Each joint force component, however, has unique needs that are recognized by the OSD and the CJCS.[10]

---

[10]The CPR document (fall 1994), which is sent from the CJCS to the Secretary of Defense (SECDEF), noted that command, control, communications, computers, and intelligence (C^4I) was a critical element in the support of the joint force component. The CPR indicated that the information

All the services provide space-based assets. The Air Force provides multiple capabilities: heavy launch, satellite positioning, etc. The Air Force has defined a core capability called **space superiority**; its goal is to dominate in the near future the critical strategic decisionmaking and resourcing in this area. Therefore, how might the Air Force ensure that providing space-based military capabilities is viewed as part of its space superiority core competency/capability?

This assessment has concentrated on how the demand is shaped when attempting to look beyond the program years. The discussion focuses on a number of interrelated elements: the ongoing assessment of alternative futures, which baselines potential critical changes in the strategic environment and identifies future tasks that might need to be performed; the Air Force's vision, which defines the goals and objectives of the Air Force 15 to 25 years out; and finally, the use of core competencies/capabilities as a way to define and refine an organization.

---

demands for the various component commanders varied. The Air Force, for example, needs data on fixed targets. Often the Army is responsible for defining mobile target sets.

4. Defining the Supply

Our definition of new-concept development includes both materiel and nonmateriel innovation. Innovative ideas, therefore, cover the spectrum— science and technology, policy, operational concepts, new organizational and training processes, and doctrine. Historically the Air Force has focused on technological solutions.[1] Only recently has the Air Force attempted to establish a strong doctrinal capability that links its operational concepts and training to doctrine. The current and projected defense fiscal environment suggests that the Air Force needs to identify, assess, and integrate both materiel and nonmateriel solutions (within a joint context) to sustain itself over time as an institution. These kinds of ideas form the supply side of new-concept development.

Internal to the Air Force are a number of areas that foster materiel and nonmateriel innovative ideas. These include the laboratories, major commands (MAJCOMs), Headquarters and Air Staffs, and Air University, as well as lessons learned from operations and exercises.

The external community, however, is also a source for new ideas and concepts. These include public and private universities, the technology-based federally funded research and development centers (FFRDCs), foreign governments, the U.S. government, the other military departments, publications, professional meetings, and commercial ventures.

Changes in the Innovation Environment

When looking at the traditional sources of innovative ideas in the United States, some discussion of that environment and its ability to provide new concepts is required. Historically, the military services have dominated the federal research and development (R&D) budgets. The focus of these activities has been on the development of high-technology capabilities, their testing, and ultimately, their incorporation into the inventory. DoD R&D money was often invested in commercially based ventures focused on defining innovative capabilities for the DoD. In the 1980s, R&D investment focused on technology development (not on

---

[1]The relatively new Mission Area Planning System (MAPS) process focuses almost exclusively on the modernization choices of the MAJCOMs and their prioritization within the MAJCOMs and ultimately by the Air Force.

nonmateriel solutions), with most investment concentrated primarily in two areas: aircraft and missile development.[2]

Since the early 1990s, however, in response to declines in the defense budget, the services have used R&D and procurement dollars as "bill payers" for sustaining force structure and existing inventory.[3] This has caused a falloff in the identification of new technologies and the development of weapon systems (Mowery and Rosenbrier, 1993). The impact of the decline of DoD funding has yet to be fully assessed, but one known repercussion is that the services are forced to seek more nonmateriel solutions to meet their capability demands. In the area of nonmateriel solutions, the Air Force does not have a strong institutional underpinning. The Air Force concluded early in its institutional history that its survival depended on high technology. Once an aircraft was developed, the organizational support—training, doctrine, sustainment—would follow. This approach differs substantially from that of such organizations as the Army, which views its institutional underpinning as resting on in its doctrine and its ability to develop and justify all systems based on its doctrine and training.[4] The dilemma, which will be discussed later, is that declines and shifts in DoD investment have led to a breakdown in the traditional military-industrial relationship.

Since the conclusion of World War II, U.S. universities and FFRDCs have emerged as international centers for technological innovation. Federal expenditures to these institutions have been in the form of contracts and grants for specified research areas. Most of the "demand" for scientific research has been directed by federal departments or agencies, which often have different responsibilities and goals (Nelson, 1993, p. 48).

The ties between the federal government and universities in encouraging technological and scientific innovation have also fostered the education and training of scientists, technologists, and engineers. For instance, federal grants and such incentives as the G.I. Bill have enabled individuals to attend leading technology-development colleges, resulting in the push for greater research. The impacts of declines in education and research grants due to overall decreases in U.S. educational funding have yet to be determined; however, there appear to be some indications that U.S. universities are not able to sustain many of the scientific research programs that they once did.

---

[2]See DoD Budgets 1980–1985. These two areas in 1984 absorbed over 80 percent of DoD's R&D budget. (Nelson 1993) p. 43.

[3]See CBO data, presented February 7, 1996 by C. Williams.

[4]Discussions with Air Force Historian, Richard Hallion, spring 1994.

Ties between private industry and federally sponsored R&D have also played key roles in U.S. technological innovation. Small and relatively new firms are significant generators of innovation. They have been critical in the commercialization of new technologies, which is attributable, in part, to the relationships among small businesses, the university system, and government. Often, ideas developed in universities or within government agencies are transferred by individuals to the private sector, where firms are established to commercialize them.

During the late 1970s and 1980s, the competition from foreign governments, combined with changes in the telecommunications industry, challenged U.S. predominance in technological innovation. The transfer of technology overseas and the global nature of the U.S. economy seriously undercut the alliances among universities, industry, and the government.

In response to these changes, a number of new alliances have emerged. Private firms are now adopting new practices to try to exploit R&D and innovation developed outside of the firm. These include domestic and international consortia or alliances and domestic university-industry research ties (Nelson, 1993, p. 53). The federal government has also attempted to respond to these challenges by defining new ways of funding research and developing better protection for intellectual property. The government is now supporting technological research that has widespread commercial applications rather than being linked directly to the military environment.

One of the most fundamental areas of change has been university and industry research cooperation. Since the early 1990s, there has been increased collaboration between universities and industry. The motivation has been that, although U.S. educational investment is declining, university-sponsored research and technology have increasingly dominated overall U.S. research. In 1978, universities accounted for 76 percent of all combined basic research budgets of universities and industry. This number has increased in the 1990s to somewhere around 85 percent (Nelson, 1993, p. 53).

This brief assessment of where the different elements of the "good ideas" can be found points out some potential conflicts that may occur as the Air Force seeks nonmateriel innovation in response to its fiscal and mission realities. Projections for the next 15 years are that defense expenditure will decline and that new mission demands will be made on the DoD. While universities and industry are pushing the development of new technologies and scientific advances in technology, the emphasis is now on the commercial applications.

Although the Air Force needs to continue to incorporate technological innovation into its planning and procurement activities, it must also ensure that innovation includes nonmateriel solutions. This might necessitate a greater focus on the part of the Air Force, since such solutions are not major areas of concentration in industry or the universities. These innovations will probably have to come from within the Air Force, which historically has looked to technology solutions. Its core capabilities are shaped by technology areas.

Organizational Implications for the Air Force

Within the Air Force, there is no central place where ideas can be collected, screened, pursued, and/or disregarded. Currently, the MAJCOMs are chartered with identifying operational needs identified through their modernization planning process, which is based on the mission areas that they oversee. The difficulty with this organizational arrangement is that the mission areas are defined within the Air Force and do not necessarily link to the joint environment. The MAJCOMs' organizational and functional structures are designed to identify technological solutions within Air Force mission areas. Another shortcoming is that innovative ideas are usually shaped and represented within the MAJCOM stovepipes, thereby precluding the generation of larger concepts that could apply to the corporate Air Force. The Air Staff and Secretariat staffs are functionally and organizationally aligned to respond to the stovepipes rather than to integrate ideas and apply them across the entire Air Force.

5. Integration and the Operational Thread

This section discusses the integration of the demand and supply elements. The integration process contributes to resource analysis and helps define the Air Force's strategic direction and investment strategies, both of which iteratively shape the demand and supply.

The new-concept development integration function is different from the fiscally constrained planning integration function, in which one attempts to balance the total demand (requirements) against total available resources (supply). In new-concept development, integration focuses on balancing an array of ideas (supply) against a number of demand elements—alternative visions and futures, and core competencies/capabilities—that define in broad terms what needs to be supplied. The result, rather than being a fiscally constrained defense program, is refined core competencies/capabilities, alternative visions, feasible concepts, selected strategies, and investment decisions. The process is therefore iterative and at certain points must have an ability to link to any PPBS phase.

The integration function consists of a number of elements that enable it to input to long-range planning activities and, when appropriate, link to the current environment. Some of the most critical elements are to

- Articulate a vision and core competencies/capabilities within a joint operational objective and task framework
- Define the current baseline (FYDP and beyond)
- Consider the impacts of the future security environment
- Determine potential new joint operational objectives and tasks
- Assess new operational tasks against current vision and investment strategies
- Identify activities essential to the desired outcome—concepts, doctrine, technology, etc.
- Develop proposed strategies.

New-concept development is a continuous process that links across the Air Force. Figure 5.1 is a notional presentation of how the process iteratively inputs to all major Air Force activities. The left-hand side of the figure lists the organizations that are involved in fiscally constrained planning and

28

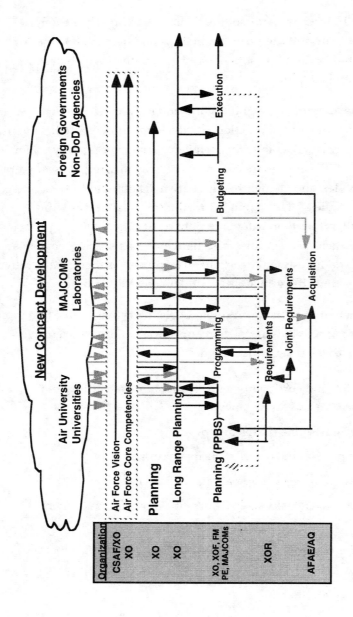

Figure 5.1—New-Concept Development Is a Continuous Process

requirements generation. The figure shows how concept development could proactively influence the Air Force in a number of areas. In our proposed process, new-concept development links throughout the Air Force and includes the identification of requirements and acquisition alternatives in both the procurement and planning areas. New-concept development is shown at the top of the figure as garnering new ideas from a variety of sources. The need for new ideas is driven by the demand side.

New-concept development also links to the PPBS process because new ideas and concepts must influence the Air Force's investment strategies. Furthermore, the Air Force was concerned with how a process, such as new-concept development, that is not fiscally constrained might link to the Air Force's PPBS deliberations. We, therefore, defined how new-concept development would link to the various phases of the PPBS process. Figure 5.2 shows our conceptualization of how this might work. The upper portion of the figure shows the idealized PPBS process. The left-hand column identifies the organizational elements by PPBS phase—the Secretary of the Air Force, the CSAF, and Plans and Programs. The lower portion of the figure shows how the outputs of new-concept development convert to the PPBS. Importantly, the lines show the process to be iterative and ongoing.

The strongest link new-concept development has to the PPBS is during the planning phases. The process interacts with the MAPS and FAPS in that it formulates and defines new concepts to assist the MAJCOMs in their development of new operational concepts and the definition of future requirements. The process also ties to long-range planning through the development, for instance, of a white paper briefing that discusses the relationships between proposed new concepts and articulated Air Force requirements. The proposed concepts are reviewed and discussed in the Air Force, and alternatives are proposed.

New-concept development could also provide information for the justification of the Air Force's program to the OSD, Joint Staff (JWCA/JROC), and Congress. Therefore, the process also has a role in the various budgeting activities that occur in the last phase of the PPBS cycle.

Illustrative Operational Thread

The gaming of new-concept development and its links to the Air Force's various resource decision processes enabled evaluation of its applicability and potential utility to the Air Force (see Figure 5.3). The "operational thread" needed to contain all the essential elements of the proposed framework and process and to demonstrate its links to ongoing Air Force strategic planning and resource

30

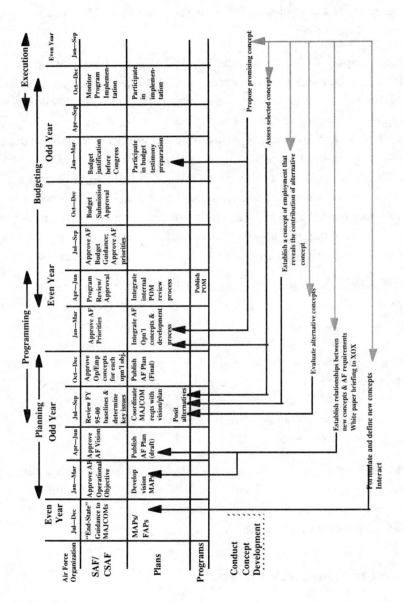

Figure 5.2—Linking Concept Development to Air Force Planning, Programming and Budgeting Process

decision processes. The space issue was identified for further evaluation (see the initial discussion of core capabilities in Section 3 of this paper). The question asked was: How does the Air Force make space superiority a core competency/capability? The operational thread contains five elements, which were individually identified as essential to process (discussed in Sections 3 and 4 of this report):

1. **Define the Current Baseline.** The current baseline identifies the demand for space-based assets to support joint operational objectives and tasks, identifies the current DoD investment in space, and determines how well the operational objectives and tasks are currently being performed to meet the national military strategy. The operational objectives and tasks are taken from the OBPRM. The current baseline is evaluated in timeframes of "now," "at the end of the FYDP," and "2010." The 15-year examination enables planners to determine the sufficiency of current operational concepts and investments to perform any tasks that might have a space dimension. For instance, in the next 10 years, a capability might be coming into the inventory that could affect some operational concepts that currently do not use space-based assets. The evaluation also reviews the output of those JWCA activities that might have applicability to space and its future use. Such documents as the National Security Strategy, the CPR, the Defense Planning Guidance (DPG), the CPA, and the Joint Vision 2010 are useful in defining a current baseline.

   The FYDP is also critical in defining the current baseline. It identifies the major elements of the U.S. military's investment in space. It also identifies the critical participants and their level of investment. For example, such information as level of investment and by whom in space launch, space architecture, NMD, TMD, and C^4I is identified. Assessing levels of investment can reveal information on who is viewed as primary owner of a capability or simply as a major stakeholder. The C^4I capability is specifically linked to the Army's Tactical Exploitation of National Capabilities (TENCAP) program. All the services are investing in how to provide a robust NMD and TMD capability; however, the Ballistic Missile Defense Office (BMDO), an OSD field operating agency (FOA), is the primary investor and often determines which service will provide a capability and its associated funding levels.

2. **Assess U.S. role and future security environment.** Section 3 discussed the analysis of alternative futures as shaping critical elements of the demand for future capabilities. The output of that assessment enables the Air Force to posit future areas for demand. This analysis not only provides some basis

Figure 5.3—Illustrative Operational Thread

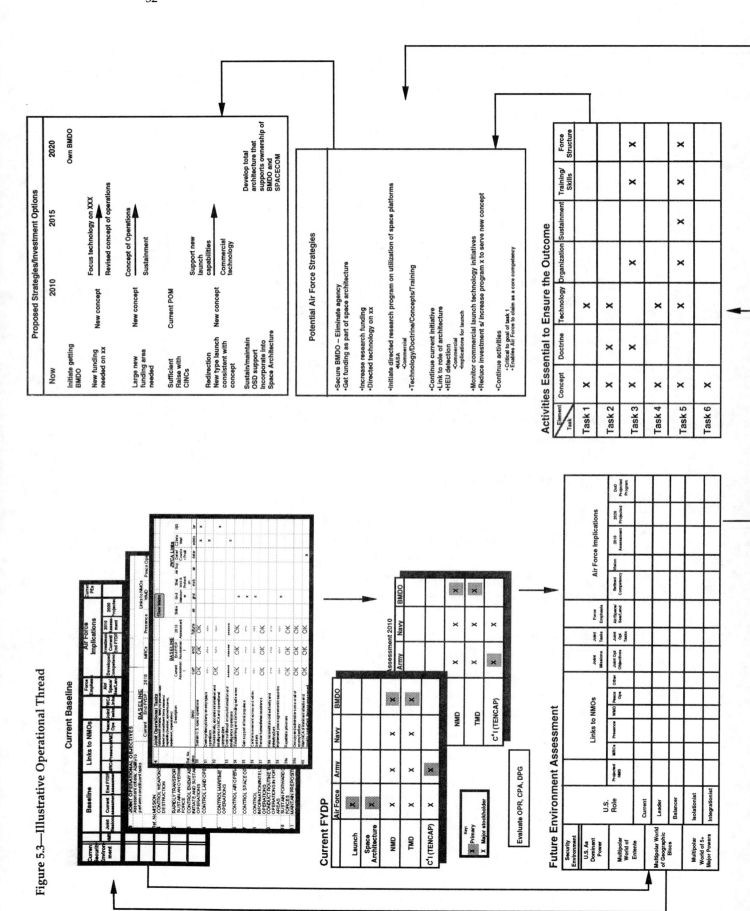

Control Space Operations ⟶ Space Superiority

Implications for Future Air Force

Joint Operational Objectives	Joint Operational Tasks	Refined Competency	Vision	2010 Assessment	2020 Projection	DoD Projected Program
Control space operations		Space superiority	Maintain space expertise / Support USSPACECOM mission	Amber/red (Current plan has significant concept and resource shortfalls)	TBD	

Disaggregate Into Sets of Operational Tasks

Possible New Joint Operational Tasks

Task 1 — Support and protect U.S. against space-based capabilities from hostile nations and terrorists

Task 2 — Develop space-based weaponry to protect nation

Task 3 — Provide and man space platforms

Task 4 — Prevent proliferation of WMD using space

Task 5 — Develop and maintain launch capability

Task 6 — Design space architecture

Canvas Internal Community
- Papers
- Roundtables
- DSB/AFSB

Canvas External Community
- Papers
- Publications
- Forums
- Etc.

Assessment Summary: Current Status

Task 1 Support and protect U.S.
- BMDO
- No overarching architecture
- Future funding looks stable

Task 2 Develop space based weaponry
- OSD focused on microwave technology
- Air Force stealth program investment
- No commercial work
- Foreign government: XXX

Task 3 Provide and man space platforms
- Foreign governments: XX
- Commercial investment (foreign substantial)

Task 4 Assure freedom of access and prevent proliferation
- No U.S. initiatives
- Key JWCA concern
- Foreign government initiatives XXX

Task 5 Capable launch
- Key Air Force competency - critical investment
- US and foreign commercial ventures
- Discussions of commercial capabilities
- New techniques could argue new type of launch capability

Task 6 Space Architecture
- Key Air Force initiative but not sufficiently organized
- OSD support

Current DoD Investment/Research

	Task 1 Current	Task 1 Projected	Task 2 Current	Task 2 Projected	Task 3 Current	Task 3 Projected	Task 4 Current	Task 4 Projected	Task 5 Current	Task 5 Projected	Task 6 Current	Task 6 Projected
Air Force	X	?	X	?		?		?	X	?	X	?
Navy	X								X		X	
Army	X										X	
OSD	X				X		X					

Key: X Primary, X Major stockholder

Also Consider Non-DoD Funded Research and Potential Competitive Space-Based Activities

	Task 1 Current	Task 1 Projected	Task 2 Current	Task 2 Projected	Task 3 Current	Task 3 Projected	Task 4 Current	Task 4 Projected	Task 5 Current	Task 5 Projected	Task 6 Current	Task 6 Projected
U.S. Based												
Universities	X	X									X	X
NASA					X	X			X	X	X	X
Commercial			X	?			X	?	X	X	X (CIA)	X
Other											X	X
Foreign												
Governments			X	X	X	X	X	X	X	X	X	X
Universities									X	X		
Commercial												
Other												

Key: X Primary, X Major stockholder

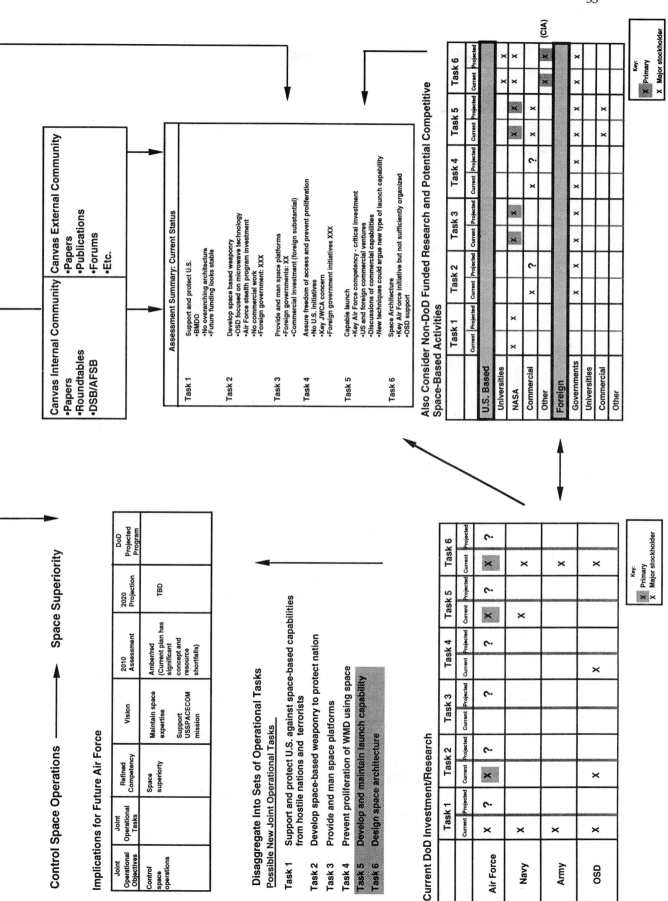

for defining new mission area, but also their associated operational objectives and tasks. Implications for the future Air Force can be extracted from such a systematic activity. Some of the implications will drawn from examining and refining core competencies/capabilities, rethinking and adjusting the institution's vision, and assessing how changed or new mission areas could affect Air Force corporate strategies in the outyears. Integral to this phase of the process is an evaluation of the implications for projected Air Force investment in relation to total DoD investment. For instance, in assessing space-based activities, the Air Force would need to examine projected Air Force and total DoD expenditures in this and related areas.

3. **Identify future operational objectives and tasks to determine need.** The output of an assessment of future environments could yield substantial information that, when linked together, would begin to shape courses of action on which to posture the future Air Force. For example, the analysis might reveal a future joint operational objective to be **control of space operations**. The operational objective could support an Air Force refined core capability of **provide space-based capability through space superiority**. The desire to develop a core capability as the provider of space capability necessitates that, in the Air Force vision, an institutional goal should be to maintain and expand its role as a provider of space-based capabilities. Part of the Air Force's planning process to underpin this goal is to support U.S. Space Command's missions. The Air Force's assessment of its planning horizon in 2010 indicates that its current operational and organizational concepts and investment strategies are insufficient to meet its institutional goals of space superiority. For instance, its near-term and 2010 R&D and procurement budgets are primarily focused on its long-range bomber programs, leaving little room for other investment considerations. Its goal, therefore, might not be attainable until the 2020 period, when small, interrelated R&D efforts and new-concept work might yield some usable results.

If the Air Force concludes that it must begin now to shape the future space environment (by approximately 2020) with small, directed investments, it needs to narrow down the defined operational objective into some sets of plausible operational tasks. Again, the Air Force needs to direct its activities toward those areas that will provide it with the greatest ability to be a critical player in the strategic decisionmaking that provides the total capability. Toward this end, we identified several possible **joint operational tasks**. The tasks do not suggest an operational concept, for part of the Air Force's concept development activities is to develop and analyze possible new operational concepts in support of the Air Force's goal to be the key provider space-based capabilities in the 2020 timeframe. Some illustrative tasks might be the following:

> Task 1. Support and protect the United States against space-based capabilities from hostile nations and terrorists.
>
> Task 2. Develop space-based weaponry to protect the nation.
>
> Task 3. Provide and man space platforms.
>
> Task 4. Prevent proliferation of WMD using space.
>
> Task 5. Develop and maintain launch capability.
>
> Task 6. Design space architecture.

The definition of the future tasks completes the activities that occur within the demand function, as discussed in Section 4. The Air Force then would assess its ability to supply new ideas and investment strategies to meet the demand. The Air Force would begin by evaluating the demand against those areas in which it is currently engaged in space-related activities. The analysis would then review current DoD investment and research in each of the task areas. For instance, work related to future Task 1 is currently being funded through the NMD and TMD programs in which all the services are participating. Future activity in this area is uncertain, given that an NMD architecture remains to be defined and agreed upon by the OSD and Congress. The Air Force is aggressively attempting to secure for itself a critical role in the definition and support of the NMD and TMD capabilities.

4. **Identify areas in which new concepts might be useful.** The task assessment could show that the Air Force is the primary player in three related areas: providing heavy launch capabilities, satellite positioning, and space architecture. It currently is the major investor in heavy launch capabilities, and it participates in both the NMD and TMD programs. The future operational objectives and tasks, however, suggest that the Air Force might want to seek new and different types of capabilities. For example, the Air Force might want to posit new launch capabilities that focus on the ability of the United States to launch and position small satellites simultaneously, which would take advantage of miniaturization technologies and emerging commercial launch capabilities. The utilization of these technological advancements would necessitate new operational concepts that would move the United States away from its strong reliance on heavy launch capabilities to smaller, reusable satellite capabilities. To advance this concept, the Air Force would investigate new launch and satellite technologies and their applicability to projected missions.

In each of the future task areas, the Air Force might want to evaluate what new technologies and concepts are being devised outside of the DoD. It could turn to U.S. universities, research centers, and federally funded agencies such as the National Aeronautics and Space Administration (NASA). Again, the evaluation would focus on where research monies are being spent and how the Air Force might incorporate these activities into the development of new concepts. Similarly, the Air Force would also assess what foreign governments, consortiums, and universities are doing in the task areas. As noted earlier, U.S. and foreign innovation is focused primarily on technological advances. If the Air Force is seeking nonmateriel solutions, it must generate major portions of them internally.

The analysis of demand and supply is integrated through an assessment summary. Each task is discussed in terms of what is occurring now and in the near future in the Air Force, DoD, and externally. The assessment summary could identify key internal or external organizations that play major roles in strategic decisionmaking in the area, underdeveloped concepts, and needed new materiel and nonmateriel concepts.

Once areas are identified for future concept development, the Air Force needs to canvas the internal and external communities for ideas. The ideas can come through papers, publications, roundtables, science boards, forums, etc. The ideas are culled based on where the Air Force initially believes solutions can be found that will provide the greatest advantage. These can include nonmateriel solutions, such as operational concepts, doctrine, organization, sustainment, and structure. Proposed concepts, however, might or should include multiple, interrelated areas.

5. **Propose alternative strategies and courses of action.** The output of the process is the identification of potential Air Force strategies. Proposed strategies or courses of action could include initiating research programs in a variety of materiel and nonmateriel areas. They could also include outsourcing some current activities over the next 15 to 20 years. For instance, the Air Force might decide to outsource a lot of its heavy launch capabilities to commercial firms and concentrate its R&D monies on developing concepts to enable the Air Force to launch clusters of satellites simultaneously. Another strategy might be to monitor and integrate (as appropriate) commercially developed capabilities.

The proposed strategies result in their presentation and debate among the Air Force leadership prior to a course of action being adopted. The strategies could call for a synergy of activities, such as technology development, redirection of current activities, and the need for new concepts. Any agreed-upon strategy must look out 15 to 20 years. An adopted strategy is not immutable; rather, it defines a course of action based on current circumstances and defined goals for the future. If circumstances change, the strategy should be examined and changed as appropriate. It is at these junctures that new-concept development, long-range planning, and PPBS activities are linked and mutually supportive.

6. Conclusions

This report has proposed a framework and process for new-concept development. The process was defined using the simple economic model of demand, supply, and integration. The report discussed the elements that shape the demand when attempting to define strategic direction and potential investment strategies in the 15- to 20-year time horizon. These include assessments of alternative futures, incorporation of the Air Force's vision, and the evaluation and refinement (when appropriate) of core competencies/ capabilities. Each of these elements, however, needs to be defined within the context of the joint operational objectives and tasks. It is only here that future capability shortfalls can be defined and comparative worth of a concept evaluated.

Historically, innovation within the Air Force has focused on technological improvements. New technologies have most often come from private industry's research efforts, which the DoD funded through R&D. Since the early 1980s, however, the declines in federal R&D and defense procurement resources have resulted in private industry refocusing its technology research on commercial ventures. As a result of this reorientation, DoD no longer always invents new ideas and has them adopted by the commercial sector; rather, DoD is now adopting technologies developed for commercial applications.

Another potential difficulty in the supplying of new ideas is that, historically, the Air Force has not focused on nonmateriel solutions. The emphasis of the modernization and mission-area analysis processes is on technology improvements, usually articulated in the requirement for a new aircraft. If nonmateriel solutions are proposed, they usually are identified and implemented within a particular MAJCOM. The stovepiped nature of this activity eliminates the potential for a nonmateriel improvement to be shared and possibly adopted throughout the Air Force.

The integration process in new-concept development is the mechanism that filters new ideas against the demand for them. It also enables the Air Force to link new concepts to its resource investment processes, such as PPBS and acquisition. The integration process, therefore, provides the handshake among a number of activities, some fiscally constrained and others not.

The illustrative operational thread of Figure 5.3 shows the interaction among the different subelements, as well as the interrelationships among the demand, supply, and integration functions. The thread attempts to functionally demonstrate how new-concept development has the potential to interact with all the major planning and resourcing processes within the Air Force. The structure enables the Air Force to identify and examine proposed concepts within a joint context and concurrently assesses and redefines core capabilities.

Air Force Long-Range Planning has agreed that this work is valuable to the Air Force. New-concept development's linkages to the planning and resourcing processes within the Air Force could be examined in greater detail. An assessment could be made to examine how the Air Force might institutionalize this process and its long-range planning activities both organizationally and functionally. Some of the issues that should be addressed are:

- How might proposed new concepts be identified as useful?

- How should new-concept development and long-range planning be functionally and organizationally supported?

- How might new-concept development and long-range planning be implemented and sustained?

Bibliography

Argyris, Chris, "Good Communication That Blocks Learning," *Harvard Business Review*, July–August 1994, pp. 77–85.

Aspin, Les, *Report on the Bottom-Up Review*, October 1993.

Baldridge, J. Victor, and Robert A. Burnham, "Organizational Innovation: Individual, Organizational, and Environmental Impacts," *Administrative Science Quarterly*, Vol. 20, June 1975, pp. 165–176.

Bansler, Jorgen P., and Keld Bodker, "A Reappraisal of Structured Analysis: Design in an Organizational Context," *ACM Transactions on Information Systems*, Vol. 11, No. 2, April 1993, pp. 165–193.

Beer, Michael, Russell Eisenstat, and Bert Spector, "Why Change Programs Don't Produce Change," *Harvard Business Review*, November–December 1990, pp. 158–166.

Blau, Judith R., and William McKinley, "Ideas, Complexity, and Innovation," *Administrative Science Quarterly*, Vol. 24, June 1979, pp. 200–219.

Crocker, Keith J., and Kenneth J. Reynolds, "The Efficiency of Incomplete Contracts: An Empirical Analysis of Air Force Engine Procurement," *RAND Journal of Economics*, Vol. 24, No. 1, Spring 1993, pp. 126–146.

Curtis, Bill, Marc I. Kellner, and Jim Over, "Process Modeling," *Communications of the ACM*," Vol. 35, No. 9, September, 1992, pp. 75–90.

Daft, Richard L., "A Dual-Core Model of Organizational Innovation," *Academy of Management Journal*, Vol. 21, No. 2, 1978, pp. 193–210.

Daft, Richard L., "Bureaucratic Versus Nonbureaucratic Structure and the Process of Innovation and Change," *Research in the Sociology of Organizations*, Vol. 1, 1982, pp. 129–166.

Damanpour, Fariborz, "The Adoption of Technological, Administrative, and Ancillary Innovations: Impact of Organizational Factors," *Journal of Management*, Vol. 13, No. 4, 1987, pp. 675–688.

Damanpour, Fariborz, and William M. Evan, "Organizational Lag," *Administrative Science Quarterly*, Vol. 29, 1984, pp. 392–409.

Davis, Tom, "Effective Supply Chain Management," *Sloan Management Review*, Summer 1993, pp. 35–46.

Dewar, Robert D., and Jane E. Dutton, "The Adoption of Radical and Incremental Innovations: An Empirical Analysis," *Management Science*, Vol. 32, No. 11, November 1986, pp. 1422–1433.

Ettlie, John E., William P. Bridges, and Robert D. O'Keefe, "Organization Strategy and Structural Differences for Radical Versus Incremental Innovation," *Management Science*, Vol. 30, No. 6, June 1984, pp. 682–701.

Gennell, Mary L., "Synergy, Influence, and Information in the Adoption of Administrative Innovations," *Academy of Management Journal*, Vol. 27, No. 1, 1984, pp. 113–129.

Goodman, Paul S., and James W. Dean, Jr., "Creating Long-Term Organizational Change," in *Change in Organizations*, San Francisco: Jossey-Bass Publishers, 1982, pp. 226–279.

Grinyer, Peter H., and Masoud Yasai-Ardekani, "Strategy, Structure, Size and Bureaucracy," *Academy of Management Journal*, Vol. 24, No. 3, 1981, pp. 471–486.

Huber, George P., "A Theory of the Effects of Advanced Information Technologies on Organizational Design, Intelligence, and Decision Making," *Academy of Management Review*, Vol. 15. no. 1, 1990, pp. 47–71.

Khalilzad, Zalmay, ed., *Strategic Appraisal 1996*, Santa Monica, Calif.: RAND, MR-543-AF, 1996.

Kimberly, John R., and Michael J. Evanisko, "Organizational Innovation: The Influence of Individual, Organizational, and Contextual Factors on Hospital Adoption of Technological and Administrative Innovations," *Academy of Management Journal*, Vol. 24, No. 4, 1981, pp. 689–713.

Krass, Iosif A., Mustafa C. Pinar, Theodore J. Thompson, and Stavros A. Zenios, "A Network Model to Maximize Navy Personnel Readiness and Its Solution," *Management Science*, Vol. 40, No. 5, May 1994, pp. 647–661.

Lee, Hau L., and Corey Billington, "Managing Supply Chain Inventory: Pitfalls and Opportunities," *Sloan Management Review*, Spring 1992, pp. 65–73.

Lee, Hau L., and Corey Billington, "Material Management in Decentralized Supply Chains," *Operations Research*, Vol. 41, No. 5, September–October 1993, pp. 835–847.

Lewis, Leslie, Preston Niblack, William Schwabe, and John Schrader, *Overseas Presence: Joint Warfighting Capabilities Assessment Conference*, Santa Monica, Calif.: RAND, PM-359-JS, 1995.

Lewis, Leslie, James A. Coggin, and C. Robert Roll, *The United States Special Operations Command Resource Management Process*, Santa Monica, Calif.: RAND, MR-445-A/SOCOM, 1994.

Malone, Thomas W., "Modeling Coordination in Organizations and Markets," *Management Science*, Vol. 33, No. 10, October 1987, pp. 1317–1332.

Malone, Thomas, W. and John F. Rockart, "Computers, Networks and the Corporation," *Scientific American*, September 1991, pp. 128–136.

Mazmanian, Daniel A., and Paul A. Sabatier, *Effective Policy Implementation*, Lexington, Mass.: Lexington Books, 1981.

Meyer, Alan D., and James B. Goes, "Organizational Assimilation of Innovations: A Multilevel Contextual Analysis," *Academy of Management Journal*, Vol. 31, No. 4, 1988, pp. 897–923.

Mowery, David C., and Nathan Rosenbrier, "The U.S. National Innovation System," in Nelson (1993), pp. 46–48.

Nelson, Richard R., ed., *National Innovation Systems: A Comparative Analysis*, New York: Oxford University Press, 1993.

Pentland, Brian T., and Henry H. Rueter, "Organizational Routines as Grammars of Action," *Administrative Science Quarterly*, Vol. 39, 1994, pp. 484–510.

Pirnie, Bruce, *An Objectives-Based Approach to Military Campaign Analysis*, Santa Monica, Calif.: RAND, MR-656-JS, 1996.

Quinn, James Brian, *Intelligent Enterprise*, New York: The Free Press, 1992.

Rainey, Hal G., Robert W. Backoff, and Charles H. Levine, "Comparing Public and Private Organizations," *Public Administration Review*, March–April 1976, pp. 233–244.

Robert, Karlene H., Suzanne K. Stout, and Jennifer J. Halpern, "Decision Dynamics in Two High Reliability Military Organizations," *Management Science*, Vol. 40, No. 5, May 1994, pp. 614–624.

Roessner, J. David, "Incentives to Innovate in Public and Private Organizations," *Administration and Society*, Vol. 9, No. 3, November 1977, pp. 341–365.

Ross, Paul F., "Innovation Adoption by Organizations," *Personnel Psychology*, Vol. 27, 1974, pp. 21–47.

Sashkin, Marshall, and W. Warner Burke, "Organization Development in the 1980's," *Journal of Management*, Vol. 13, No. 2, 1987, pp. 393–417.

Sayles, Leonard R., *Managing Large Systems*, New Brunswick, N.J.: Transaction Books, 1993.

Smith, Hedrick, *Rethinking America*, New York: Random House, 1994.

Thompson, Victor A., "Bureaucracy and Innovation," *Administrative Science Quarterly*, May 1995, pp. 1–20.

Tichy, Noel M., Michael L. Tushman, and Charles Fombrun, "Social Network Analysis for Organizations," *Academy of Management Review*, Vol. 4, No. 4, 1979, pp. 507–519.

Tushman, Michael L., "Special Boundary Roles in the Innovation Process," *Administrative Science Quarterly*, Vol. 22, December 1977, pp. 587–605.

Williams, Cynthia, Director, Congressional Budget Office, "DoD Expenditures," briefing delivered at RAND, February 7, 1996.

Wilms, Wellford W., Alan J. Hardcastle, and Deone M. Zell, "Cultural Transformation at NUMMI," *Sloan Management Review*, Fall 1994, pp. 99–113.

Wilson, James Q., "Reinventing Public Administration," *Political Science and Politics*, December 1994, pp. 667–673.

Zald, Mayer N., "Organization Studies As a Scientific and Humanistic Enterprise: Toward a Reconceptualization of the Foundations of the Field," *Organization Science*, Vol. 4, November 1993, pp. 513–528.